D1604743

THE CRAB FROM YESTERDAY

The Life-Cycle of a Horseshoe Crab

THE CRAB FROM YESTERDAY

The Life-Cycle of a Horseshoe Crab

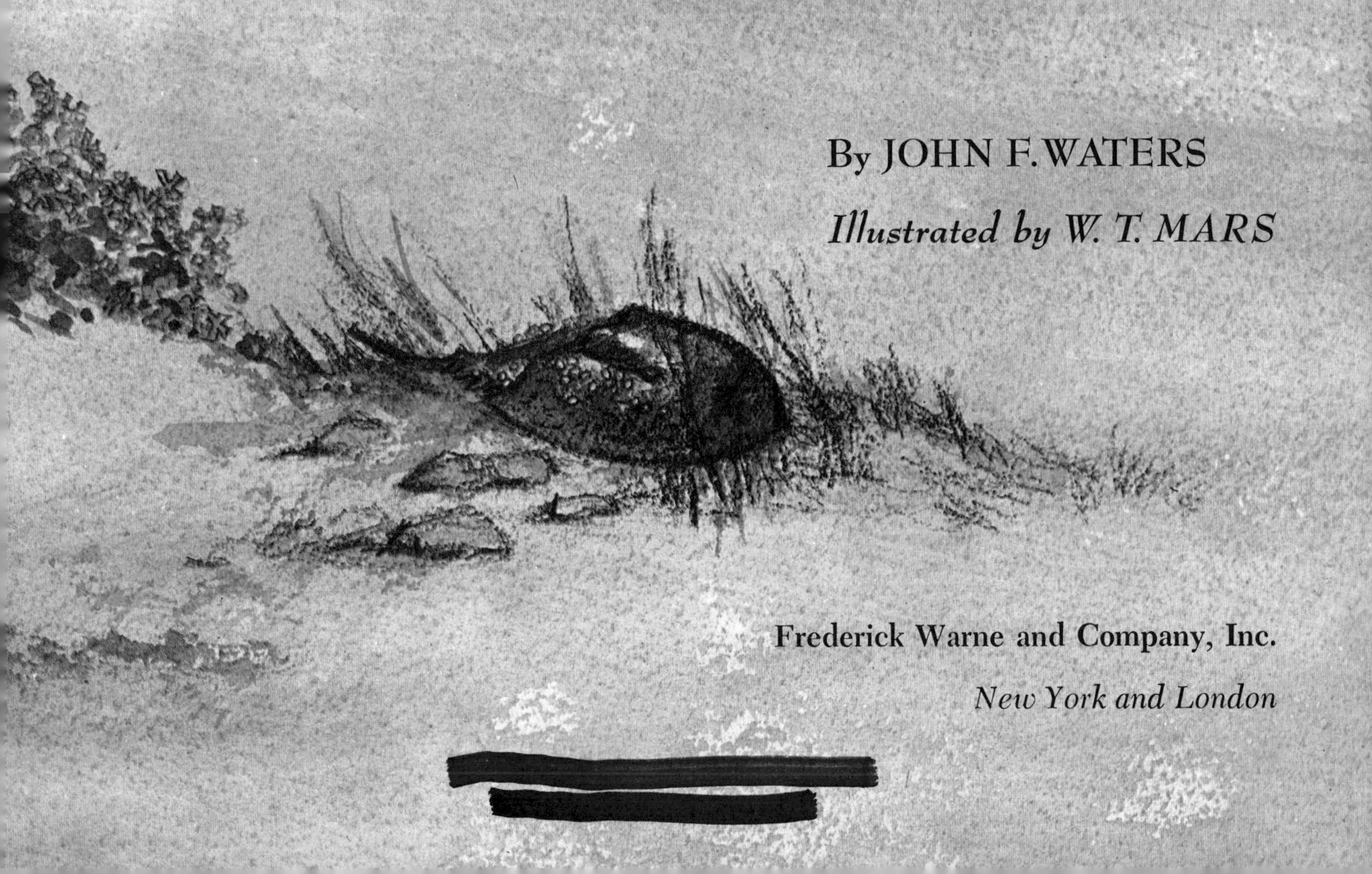

By JOHN F. WATERS

Illustrated by W. T. MARS

Frederick Warne and Company, Inc.

New York and London

Library of Congress Catalog Card Number 78-123005

Manufactured in the U.S.A.

With thanks to Carl N. Shuster, Jr.

Night blanketed the beach, and endless waves burst into a million bubbles. The spring moon was full, drawing the tides high. Birds and ducks had nested for the night, and a soft breeze brushed the beach grass into gentle waves. It was peaceful, but not quiet—for the air was filled with strange sounds. Thumping and clattering noises greeted the early dawn.

Along the edges of the saltmarshes, bays and inlets, thousands of male horseshoe crabs awaited the females. Each female was full of eggs. An inner urge was driving her, forcing her toward shallow water. It was time for her to make nests.

One female, her hard, horny shell cracked, scratched and gouged, was older than the others. The top of her shell was laced with large, white barnacles. Three boat shells clung to her under-shell along with bits of green algae. A long, spiny swordtail stretched out behind. In the moonlight, she was indeed a weird sight.

As she neared the beach, she released a jelly-like ooze. Its smell spread through the shallow water to the waiting male horseshoe crabs. They became excited and began to crawl toward the old female, bumping and thumping each other as they moved. This was the strange noise in the spring night. As she moved toward the beach, a male approached her from behind.

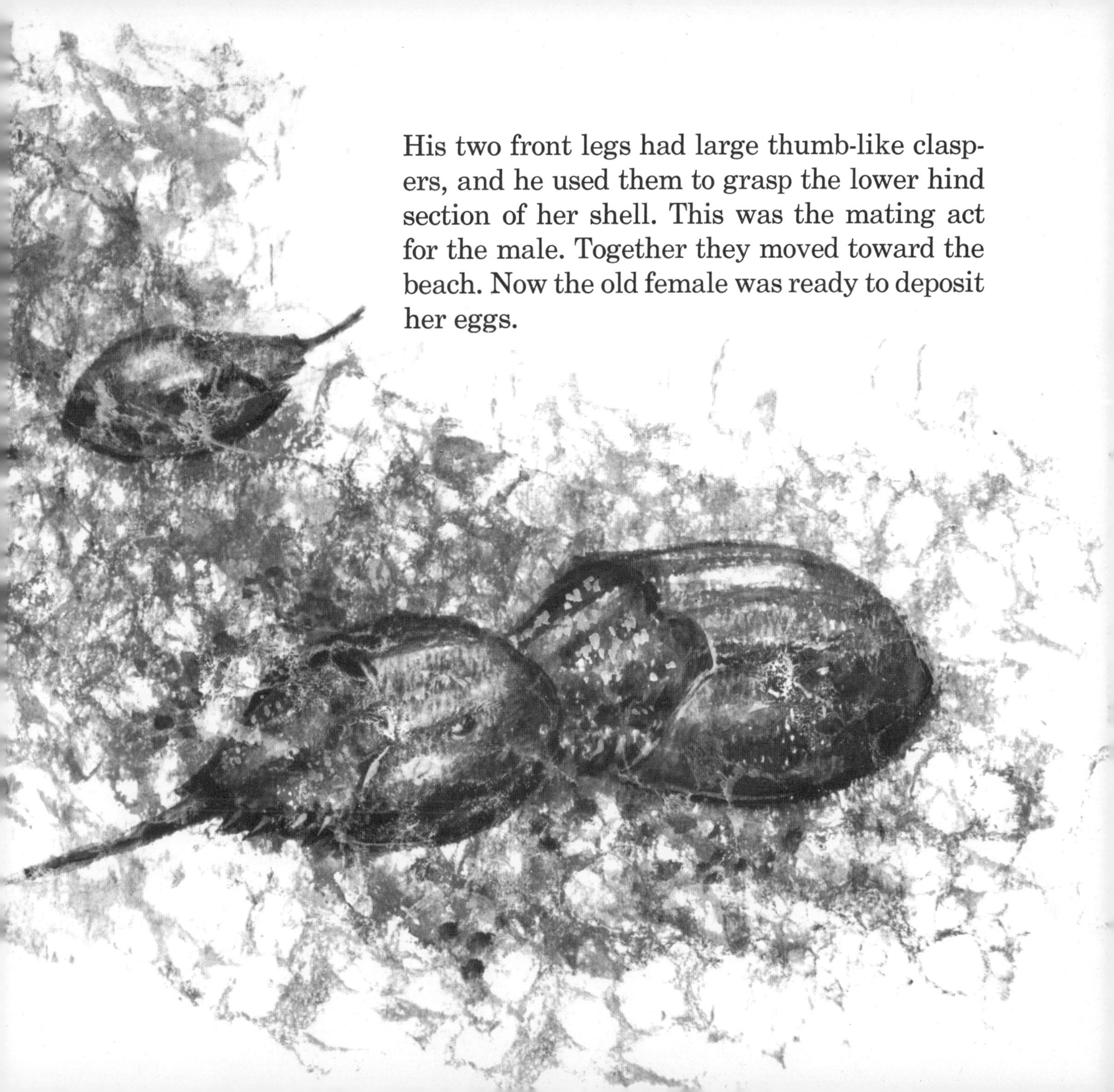

His two front legs had large thumb-like claspers, and he used them to grasp the lower hind section of her shell. This was the mating act for the male. Together they moved toward the beach. Now the old female was ready to deposit her eggs.

Where the water was only a few inches deep, the female scratched the soft, wet sand with two of her four pairs of walking legs. A large fifth pair toward the rear were used for pushing. These legs were larger and of different design, having several hinged flaps that prevented the legs from digging too deeply into mud or sand. A small sixth pair of legs in the front was to help guide food into her mouth. Once the sand had been moved aside, she snuggled down into the shallow hole and deposited several hundred eggs. They were about the size of large grains of sand—some light gray, others pale green. The "smell" of a substance mixed with the eggs was a signal for the male to fertilize them. He ejected millions of tiny sperm into the water. As a sperm found an egg, it buried itself inside. At this moment the egg became fertilized and began to develop and grow into a new baby horseshoe crab.

During the short time when the tide was highest, the old female made half a dozen nests,

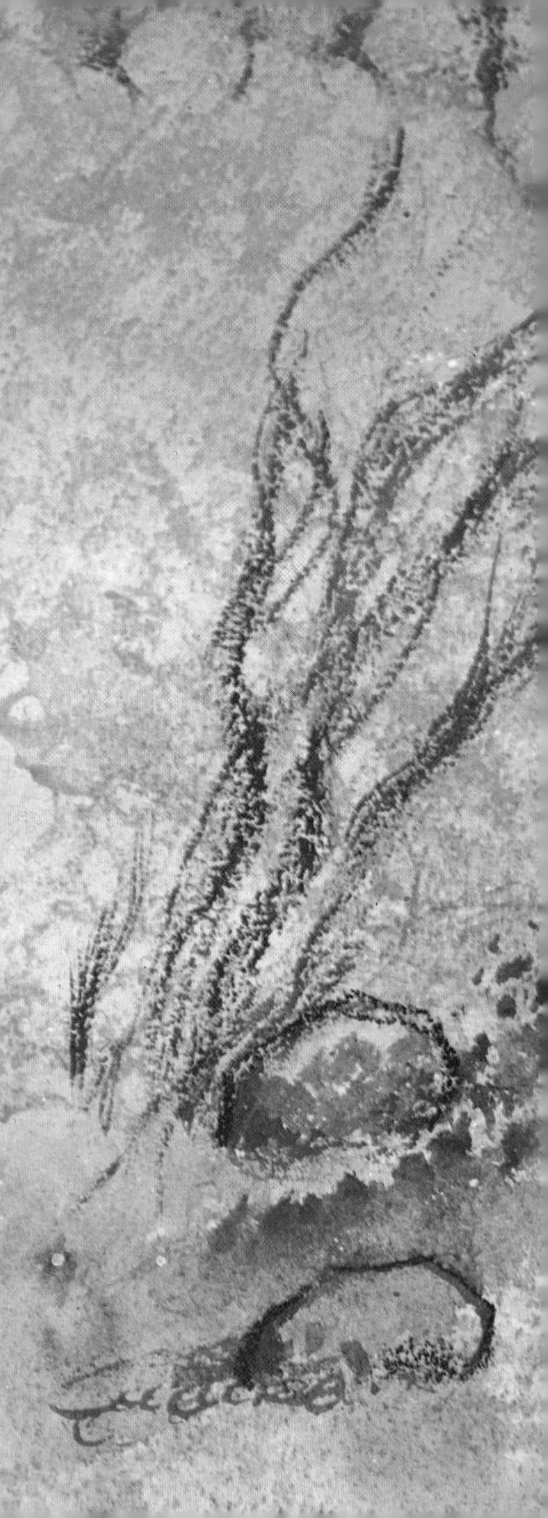

dropping about 200 eggs into each. The tide did not stay high long enough for her to finish depositing all of her 20,000 eggs. She and her mate had to go into deeper water to await the next high tide early the next morning.

After the night had passed, they returned to the beach where many other females were scratching nests in the sand. Suddenly the

dawn sky became dark. Sea gulls and terns, thick as snowflakes in a blizzard, swooped down for a feast. The birds attacked the horseshoe crabs, flipped them over and pecked at their soft underbellies. As the birds gobbled up eggs, they shrieked and squawked. Male crabs were pulled from the females. The beach was a mass of flailing wings and feet. Attracted by the scent, small fish and eels rushed to the shallow water. They went into a feeding frenzy, snapping up everything that would fit into their mouths. After a half hour, the birds flew away and the eels and fish, stuffed with eggs, swam to deeper water.

The beach was covered with broken horseshoe crabs. The old female had been pecked and scarred, her belly torn open and the eggs

ripped from her. But she was tough and still alive. She was lying upside down in a few inches of water while her male partner, also upside down, was stranded high and dry on the beach. She dug the end of her swordtail into the sand and, using it as a lever, slowly turned herself over and made for the safety of the water. The male struggled and finally righted himself, but he was confused and could not find his way back to the water. He burrowed into the sand to await the next high tide. But by then the old female would be far off. Unlike mother hens, she forgot all about her eggs once they had been laid.

The old female knew that shallow creeks in saltmarshes were fine places to hunt for food, and she enjoyed the warmer water. Later in the morning when she was burrowing in the mud of the saltmarsh, she was suddenly snatched by her swordtail and pulled from the water. A boy with brown wavy hair held her with both hands. He looked at her, counted the three boat shells fastened under her shell and examined the crack in her snout. In the middle of her underside he saw an opening at the base of her legs. It was her mouth. Pincers on her leg tips caught different kinds of clams and carried them to the bases of her legs. These "shoulders" chewed the food and forced it into her mouth. Bits of shell that could not be digested were tossed out.

On the upper part of her shell the boy noticed a pair of large eyes. They had many lenses similar to those of the common housefly. Scientists believe that the eyes can see light patterns in the sky. One pattern leads the female to land, and another leads her back to the water. In the very front of her shell, she had a small pair of eyes that sees shapes and shadows.

The boy with the brown hair held her firmly by the swordtail and laughed as he ran up the beach with his catch. A crowd of people, mostly other boys and girls, stood near a truck. The boy ran to a splintered wooden table and handed the old female to a man who tossed her onto the back of a rusted dump truck piled high with horseshoe crabs. Then he gave the boy four pennies. The boy smiled and put the pennies in his pocket. Many other children swapped a live horseshoe crab for four pennies.

They shouted and laughed as they searched the beach and saltmarsh hoping to find more horseshoe crabs to trade for pennies. This was a bounty. The children had been told that if they destroyed enough horseshoe crabs, they could save the valuable soft shell clam flats.

The boy went back to the marsh to continue the hunt. He did not know much about horseshoe crabs nor did he know that in the sand he was walking on, the old female had laid her eggs only the night before. Overnight those eggs had begun to change. They had started to swell and, in ten days, the outer covering would split open. Then each egg, about the size of a small pea, would look like a tiny crystal ball.

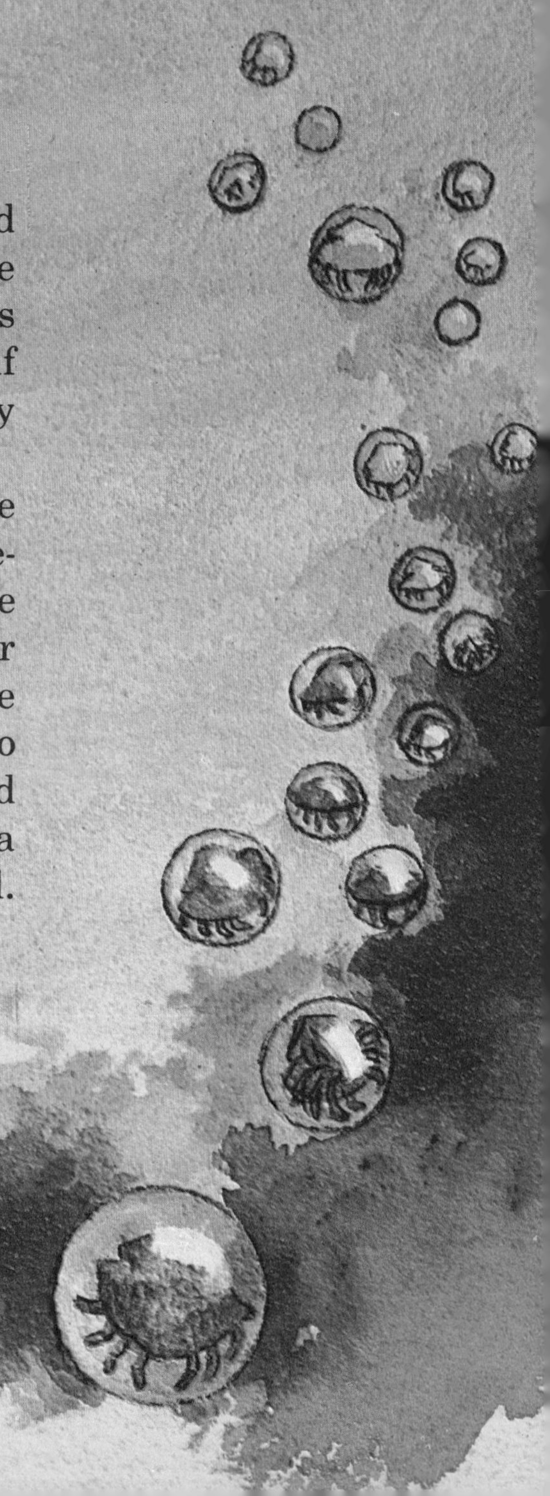

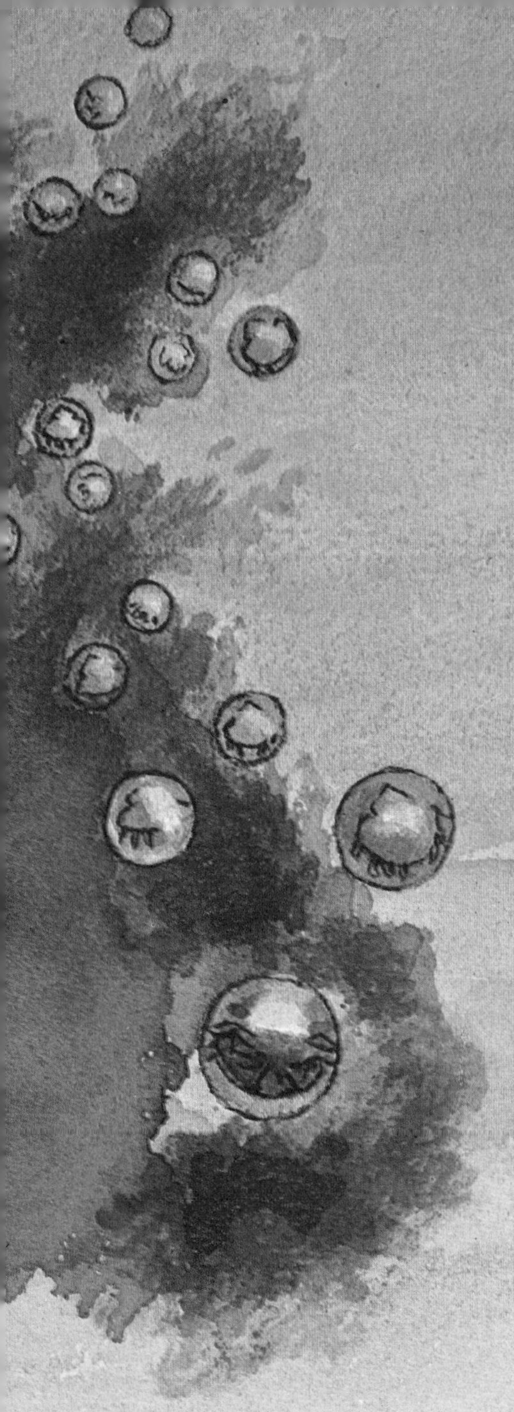

Inside would be a baby horseshoe crab complete with eyes, legs and gills, but without a swordtail. Each little horseshoe crab would be very busy turning round and round, peering at the dark world outside.

Like her children, the old female had begun her life in darkness. Many years before, when she was a tiny egg, she had waited almost a month for the full moon tides to return. Then the waves of the high tide washed away the sand that surrounded her. As her egg rubbed against the rough sand, it suddenly burst open and she escaped. She was now a larva about one quarter inch long, and she immediately began to swim. Since her shell was thick and top heavy, she swam upside down most of the time. She looked like a fat person doing the backstroke because she used her legs and gills together as paddles. Her gills looked like pages in a book and were used for breathing. As the gills beat back and forth, oxygen was taken from the water. Oxygen is a gas that all animals must have to live.

As the old female swam about as a tiny larva, she did not hunt for food. She didn't have to. She had enough food inside her body to last for many days. One day, a small puffer fish noticed her. Puffer fish learn early that baby horseshoe crabs are very tasty. The fish raced after the little horseshoe crab as she zigged and zagged in every direction. The puffer opened its mouth showing its sharp teeth. Snap! The female luckily had zigged out of the way, but the puffer fish kept up the

chase. Its mouth opened once more, but just then a clump of floating seaweed drifted by. The female bumped into it by accident, and luckily was hidden from the puffer fish. The fish looked for its lost meal, then swam off still hungry. The female had escaped from one of her many natural enemies.

That is the way it is in the sea. Animals are either eating or being eaten by one another. Baby horseshoe crabs that survive attacks from birds, fish and eels themselves grow up to eat other animals.

In order to grow, the female changed. After three weeks in the larval part of her life, she stopped swimming and settled down on the bottom of the sea. Then she molted. The joint between the top and bottom of her shell split all across the front. She crept out from inside her own shell, only this time she was a little larger. Now she had a swordtail, and for the rest of her life she would spend much more time crawling on the bottom than swimming.

She grew rapidly during her first summer. She molted many times, splitting across the front and crawling out with a new wrinkled skin one fourth larger than before. As the female took in water, her new shell began to smooth out. While her new shell was hardening, she stayed buried in the mud. This was her temporary protection from other animals such as large fish, birds and eels that would find her a good meal with such soft skin.

Even when her shell had hardened, she spent most of her time burrowing in the mud

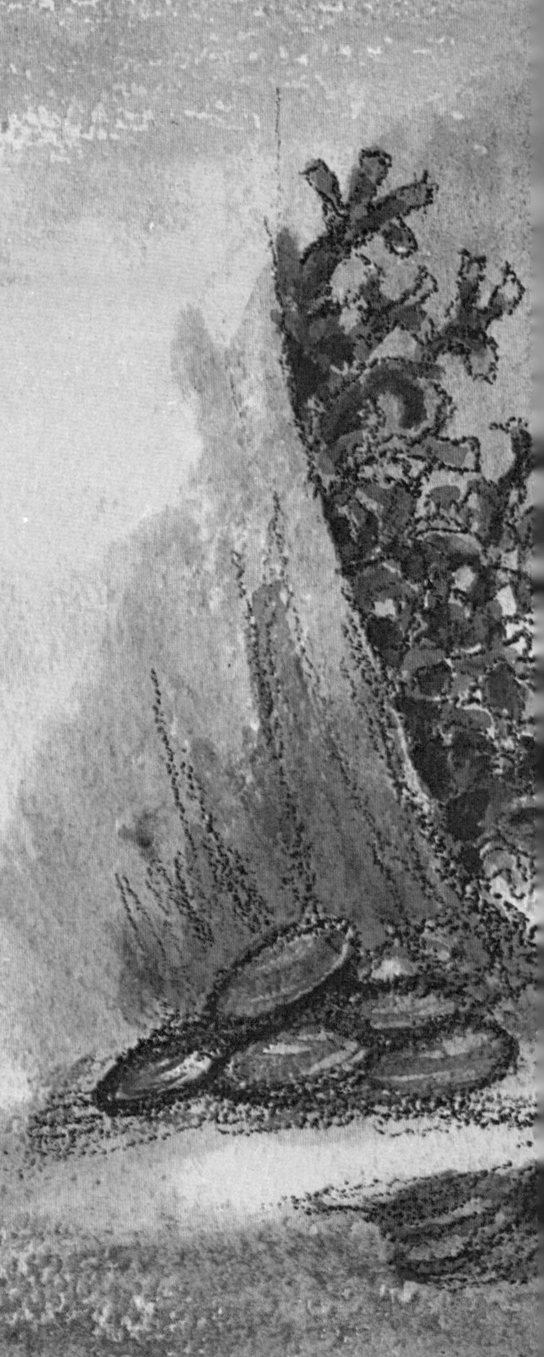

of harbors and in the sand around rocks off shore. She was always on the move, searching for small mud animals to eat. Clam worms, shellfish and tube worms were her favorites. Because she constantly plowed through the ooze like a miniature bulldozer, the top layer of the bottom mud was kept soft. Her plowing also stirred up chemicals and nutrients that mixed with the sea water and improved it for growing plants and animals.

Horseshoe crabs live near the shore with thousands of other sea animals. Some are tiny periwinkles that crawl along the rocks. Others are hard, white barnacles, black shelled mussels, five-armed starfish and pincushion sea urchins. In deeper water fuzz-covered sand dollars lie on the bottom looking like golden coins lost centuries ago from Spanish galleons.

After several years had passed, the female reached the length of an ordinary lead pencil. One day she was eating a clam when a huge male spider crab lumbered into view. He had

eight long legs and moved quickly toward the female. There was no time for escape. She burrowed into the mud. The spider crab tried to overturn her. He pushed and pulled, but the female was like a soup bowl turned upside down. She could not be budged. The spider crab gave up. He crawled away, stopping once to see if the female would come out of the mud. She did not. Her strange shape had saved her. It also keeps her from being washed off her feet by waves and from being carried long distances by strong ocean currents.

Through the years horseshoe crabs have survived most dangers. Neither birds, fish, crabs nor storms have been able to kill them off because they can live through most natural dangers. Horseshoe crabs have lived on earth for more than 300 million years. Dinosaurs have come and gone in this time.

The blue color of horseshoe crab blood is caused by a substance that carries the oxygen in the blood. Bacteria do not grow in the blood. That's why scientists have studied the blood, to see if they can find cures for diseases in man.

Called a horseshoe crab because she resembles a horse's hoof, the old female is not really a crab. She is related to spiders and scorpions. Many years ago when people in Europe first found ancestors of horseshoe crabs, they called them king crabs, and they have been called crabs ever since. Scientists call them *Limulus* because of the large eyes on their sides. *Limulus* means side-glance or look.

Years ago their fine large spiny tails were used for spear tips by Indians who hunted the eastern shores for fish and lobsters. Indians also caught them for food. Their empty shells made fine bailers for small boats. Cut in pieces, they have been used for lobster bait and pig and chicken feed.

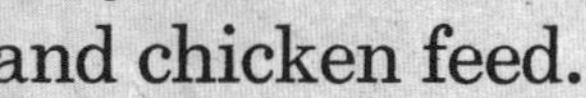

Many shellfish wardens, men who watch over the clam, scallop and oyster beds of a town or village, considered the horseshoe crab a menace. They could not see that such an ugly creature was of any use. They knew that the crabs ate many valuable soft-shelled clams, sometimes ruining whole clam beds. Scientists knew, however, that the horseshoe crab helped keep the balance of nature. It was a good animal because it ate enough worms and small clam-like animals called gem shells to keep this little shellfish from becoming a danger. They were sure that if the gem shells were allowed to grow out of control, they would crowd out other sea animals in the sand and mud flats, including the soft shelled clam.

When she was about a dozen years old, the old female molted for the last time. She stretched two and a half feet from her snout to the tip of her spiny swordtail. One day she was crawling over some rocks in shallow water. Suddenly a dark shadow passed over her.

There was a loud splash as a large rock crashed into the water and struck the old female cracking her snout. The rock stirred up the bottom mud making the water cloudy. The old female stayed hidden in the muddy water while a pair of long, black boots continued walking along the shore. On the beach were more men in boots. They held clubs and shovels. These men were helping the shellfish warden destroy as many horseshoe crabs as they could find.

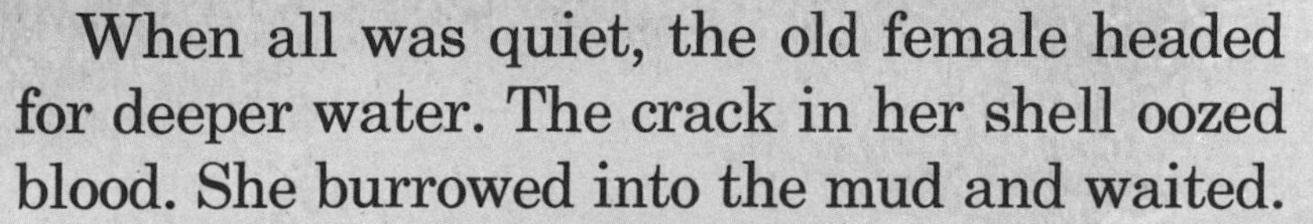

When all was quiet, the old female headed for deeper water. The crack in her shell oozed blood. She burrowed into the mud and waited.

That night a storm blew. Strong winds pushed frothy waves that crashed heavily on the beach. The female was safe in deep water and the crack in her shell had stopped bleeding.

The next day, when the morning sun rose over the sand dunes, all was peaceful. On the beach, mixed with seaweed and driftwood, were hundreds of crushed horseshoe crabs. The old female had survived another attack. But many other horseshoe crabs were not so fortunate.

In time the crack in the old female's shell healed, but it was easily noticed. She would have the crack for the rest of her life since she would molt no more. If younger, she would have lost the crack after a molt or two. But now every scrape and gouge was a permanent scar. Now, because the boy with the brown hair had caught her, the old female lay in a dump truck that was filled with hundreds of horseshoe crabs. She was buried somewhere near the middle. Already the hot sun was heating the shells of the crabs on the top of the pile. Many died on the way to the dump.

The boy with the brown hair waved at the driver as the rusty truck drove away. The

truck clanged down a dusty dirt road and approached a clearing in the woods. Shrieking seagulls flew around the heaps of rubbish and garbage. The truck backed toward the rubbish, then scattered its load of horseshoe crabs amidst the debris. Many of the horseshoe crabs were upside down, and they doubled over try-

ing to prevent their gills from drying out. If the gills dried, the horseshoe crab would die. Hundreds of swordtails stuck straight up in the air like tiny telephone poles.

The old female landed upside down at the edge of the pile. She felt the heat and dug her tail into the ground to turn herself over. Then she crawled towards the woods. Her walk was different from most animals'. She first raised her snout with her four pairs of walking legs, then she shoved herself forward with her large pair of pushing legs.

In mid afternoon, the brown haired boy came down the dusty road pulling a wagon loaded with old newspapers. The dump was in the woods, far from the saltmarsh. He placed the newspapers on a rubbish heap and was about to leave when he noticed the horseshoe crabs. Already sea birds were feasting on the remains. He walked over to the heap pulling his wagon behind him. For a while he looked at the many swordtails, then turned to leave. Then he stopped. One horseshoe crab was moving very slowly toward the woods. He picked up a stone and took aim. He was about to throw it when he saw a crack in its shell.

The boy bent down and stared at the animal. At the beach she had looked frightening; but at the dump, among broken bottles, tin cans and garbage, he saw her in a new way. She was no longer ugly and frightening. He watched as she struggled. She was so weak that she could hardly move. He dropped the stone and picked her up and turned her over. Underneath he saw the three boat shells. There was no doubt; it was the same horseshoe crab he had caught at the salt marsh. At that time, he had been very happy with his four-penny reward. Now, as he looked around and saw dead horseshoe crabs everywhere, he was not so sure.

Gently, the boy put the old female in his

wagon and started for the harbor. It was a long walk, but he wanted to go. He admired the old female's spirit as he watched her struggling in the hot sun to reach the sea even though she could not possibly get there. The boy hurried along the village streets turning now and then to look at the old female. She did not move. Perhaps it was too late. The late afternoon sun grew hotter.

Other boys, and girls too, ran up to see what was in the wagon. When they saw only an old horseshoe crab, they giggled and walked away. What would anyone want with a dead horseshoe crab?

It was still a long way to the shore. The boy's mother had told him to come right home from the dump. But somehow he had to take the poor old crab back to the sea. He wasn't sure his mother would understand.

The sun was still blazing when the boy reached the beach. The sand was so soft that he could not pull his wagon any further. In both arms he cradled the old female and carried her down to the water's edge. Gently he put her upside down in the cool water and watched to see if she moved her legs or gills. Perhaps it was too late. He watched for several minutes. There was no sign of life. The boy touched her legs. Then he saw her gills move. She *was* alive!

Many minutes went by while the brown haired boy knelt in the water and stared at the old horseshoe crab. The gills moved again. Then they began to beat faster and faster. They looked like a deck of cards being shuffled. The old female tried to dig her swordtail into the sand to turn over, but she was still very weak and could not. Carefully the boy lifted her up and turned her over. Then she moved toward deeper water where it was cooler, and there was soft mud to burrow in for worms and clams. The boy watched as the old female disappeared into the shadowy depths of the harbor.

The boy got up to leave. He looked hard for one more sign of the old female. But she was gone, perhaps never to be seen again by the boy with brown wavy hair. She was very old and could not live through another ordeal such as this.

Somehow the boy felt very good. From his pocket he pulled the four pennies, held them in his hand for a moment, then dropped them into the water, one by one. Then he too went home.